Houghton
Mifflin
Harcourt

D1515382

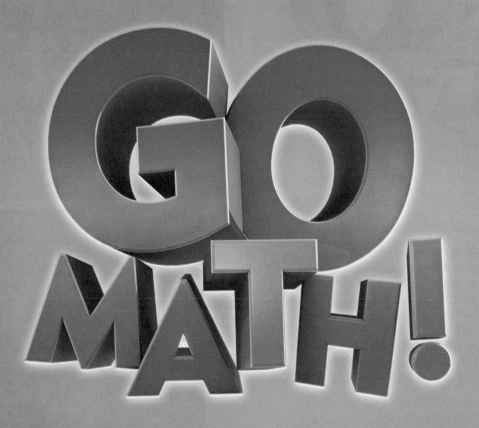

Made in the United States
Text printed on 100%
recycled paper

Houghton
Mifflin
Harcourt

Printed in the U.S.A.

ISBN 978-0-544-34247-7

13 14 15 16 17   0928   22 21 20 19 18

4500713556   B C D E F G

Dear Students and Families,

Welcome to **Go Math!**, Grade 6! In this exciting mathematics program, there are hands-on activities to do and real-world problems to solve. Best of all, you will write your ideas and answers right in your book. In **Go Math!**, writing and drawing on the pages helps you think deeply about what you are learning, and you will really understand math!

By the way, all of the pages in your **Go Math!** book are made using recycled paper. We wanted you to know that you can Go Green with **Go Math!**

Sincerely,

The Authors

Made in the United States
Text printed on 100% recycled paper

Authors

Juli K. Dixon, Ph.D.
Professor, Mathematics Education
University of Central Florida
Orlando, Florida

Edward B. Burger, Ph.D.
President, Southwestern University
Georgetown, Texas

Steven J. Leinwand
Principal Research Analyst
American Institutes for
    Research (AIR)
Washington, D.C.

Contributor

Rena Petrello
Professor, Mathematics
Moorpark College
Moorpark, California

Matthew R. Larson, Ph.D.
K-12 Curriculum Specialist for
    Mathematics
Lincoln Public Schools
Lincoln, Nebraska

Martha E. Sandoval-Martinez
Math Instructor
El Camino College
Torrance, California

English Language Learners Consultant

Elizabeth Jiménez
CEO, GEMAS Consulting
Professional Expert on English
    Learner Education
Bilingual Education and
    Dual Language
Pomona, California

Expressions and Equations

Common Core **Critical Area** Writing, interpreting, and using expressions and equations

7 Algebra: Expressions 355

COMMON CORE STATE STANDARDS

6.EE Expressions and Equations
Cluster A Apply and extend previous understandings of arithmetic to algebraic expressions.
6.EE.A.1, 6.EE.A.2a, 6.EE.A.2b, 6.EE.A.2c, 6.EE.A.3, 6.EE.A.4
Cluster B Reason about and solve one-variable equations and inequalities.
6.EE.B.6

GO DIGITAL

Go online! Your math lessons are interactive. Use *i*Tools, Animated Math Models, the Multimedia *e*Glossary, and more.

Chapter 7 Overview

In this chapter, you will explore and discover answers to the following **Essential Questions**:

- How do you write, interpret, and use algebraic expressions?
- How can you use expressions to represent real-world situations?
- How do you use the order of operations to evaluate expressions?
- How can you tell whether two expressions are equivalent?

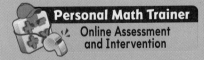

Personal Math Trainer
Online Assessment and Intervention

v

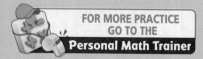

FOR MORE PRACTICE
GO TO THE
Personal Math Trainer

Practice and Homework

Lesson Check and
Spiral Review in
every lesson

Critical Area Expressions and Equations

Common Core **CRITICAL AREA** Writing, interpreting, and using expressions and equations

Great Smoky Mountains National Park is located in the states of North Carolina and Tennessee.

The Great Outdoors

The Moores are planning a family reunion in Great Smoky Mountains National Park. This park includes several campgrounds and over 800 miles of hiking trails. Some trails lead to stunning views of the park's many waterfalls.

Get Started  **WRITE** *Math*

The Moores want to camp at the park during their reunion. They will have 17 people in their group, and they want to spend no more than $100 on camping fees.

Decide how many and what type of campsites the Moores should reserve, and determine how many nights *n* the Moores can camp without going over budget. Show your work, and support your answer by writing and evaluating algebraic expressions.

Important Facts

Group Campsite
- Fee of $35 per night
- Holds up to 25 people

Individual Campsite
- Fee of $14 per night
- Holds up to 6 people

Completed by _____

Algebra: Expressions

✓ Show What You Know

Personal Math Trainer
Online Assessment
and Intervention

Check your understanding of important skills.

Name _____

▶ **Addition Properties** Find the unknown number. Tell whether you used the Identity (or Zero) Property, Commutative Property, or Associative Property of Addition. (5.OA.A.1)

1. $128 + \underline{\hspace{1cm}} = 128$

2. $(17 + 36) + 14 = 17 + (\underline{\hspace{1cm}} + 14)$

3. $23 + 15 = \underline{\hspace{1cm}} + 23$

4. $9 + (11 + 46) = (9 + \underline{\hspace{1cm}}) + 46$

▶ **Multiply with Decimals** Find the product. (5.NBT.B.7)

5. 1.5×7

6. 5.83×6

7. 3.7×0.8

8. 0.27×0.9

▶ **Use Parentheses** Identify which operation to do first. Then, find the value of the expression. (5.NBT.B.6)

9. $5 \times (3 + 6)$ _____

10. $(24 \div 3) - 2$ _____

11. $40 \div (20 - 16)$ _____

12. $(7 \times 6) + 5$ _____

Greg just moved into an old house and found a mysterious trunk in the attic. The lock on the trunk has a dial numbered 1 to 60. Greg found the note shown at right lying near the trunk. Help Greg figure out the three numbers needed to open the lock.

Lock Combination
Top Secret!

1st number: $3x$

2nd number: $5x - 1$

3rd number: $x^2 + 4$

Hint: $x = 6$

Vocabulary Builder

▶ **Visualize It** •

Sort the review words into the bubble map.

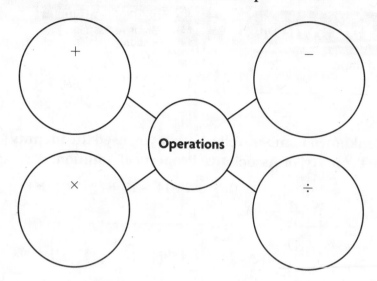

▶ **Understand Vocabulary** •

Complete the sentences using the preview words.

1. An exponent is a number that tells how many times a(n)

   _____ is used as a factor.

2. In the expression 4*a*, the number 4 is a(n)

   _____ .

3. To _____ an expression, substitute numbers

   for the variables in the expression.

4. A mathematical phrase that uses only numbers and operation

   symbols is a(n) _____ .

5. A letter or symbol that stands for one or more numbers is a(n)

   _____ .

6. The parts of an expression that are separated by an addition

   or subtraction sign are the _____ of the

   expression.

GO DIGITAL
• Interactive Student Edition
• Multimedia eGlossary

Chapter 7 Vocabulary

algebraic expression

expresión algebraica

3

base (arithmetic)

base

6

coefficient

coeficiente

10

difference

diferencia

22

equivalent expressions

expresiones equivalentes

28

evaluate

evaluar

31

exponent

exponente

32

factor

factor

33

A number used as a repeated factor

$$\underbrace{5 \times 5 \times 5}_{\text{3 repeated factors}} = 5^{3} \xleftarrow{} \text{exponent}$$

base

An expression that includes at least one variable

$$x + 10 \qquad 3 \times y \qquad 3 \times (a + 4)$$

variable · variable · variable

The answer to a subtraction problem

$$7 - 3 = 4 \xleftarrow{} \text{difference}$$

A number that is multiplied by a variable

Example: $4k$ The coefficient of the term $4k$ is 4.

To find the value of a numerical or algebraic expression

Example: Evaluating $2x + 3y$ when $x = 1$ and $y = 4$ gives $2(1) + 3(4) = 2 + 12 = 14$.

Expressions that are equal to each other for any values of their variables

Example: $2x + 4x = 6x$

A number multiplied by another number to find a product

$$2 \times 3 \times 7 = 42$$

factors

A number that shows how many times the base is used as a factor

$$\underbrace{5 \times 5 \times 5}_{\text{3 repeated factors}} = 5^{3} \xleftarrow{} \text{exponent}$$

base

like terms

términos semejantes

49

numerical expression

expresión numérica

67

order of operations

orden de las operaciones

69

product

producto

80

quotient

cociente

84

sum

suma o total

99

terms

términos

101

variable

variable

105

A mathematical phrase that uses only numbers and operation signs

$$3 + 16 \times 2^2 \qquad 4 \times (8 + 5) \qquad 2^3 + 4$$

Expressions that have the same variable with the same exponent

Algebraic Expression	Terms	Like Terms
$5x + 3y - 2x$	$5x$, $3y$, and $2x$	$5x$ and $2x$
$8z^2 + 4z + 12z^2$	$8z^2$, $4z$, and $12z^2$	$8z^2$ and $12z^2$
$15 - 3x + 5$	15, $3x$, and 5	15 and 5

The answer to a multiplication problem

$$3 \times 4 = 12 \longleftarrow \text{product}$$

A set of rules which gives the order in which calculations are done in an expression

Order of Operations
1. Perform operations in parentheses.
2. Find the values of numbers with exponents.
3. Multiply and divide from left to right.
4. Add and subtract from left to right.

The answer to an addition problem

$$2 + 5 = 7 \longleftarrow \text{sum}$$

The number that results from dividing

$$80 \div 4 = 20 \qquad 4\overline{)80} \longleftarrow \text{quotient}$$

quotient

A letter or symbol that stands for an unknown number or numbers

$$x + 10 \qquad 3 \times y \qquad 3 \times (a + 4)$$

variable   variable   variable

The parts of an expression that are separated by an addition or subtraction sign

Example:

$4k + 5$ The expression has two terms, $4k$ and 5.

Going Down the Blue Ridge Parkway

For 2 to 4 players

Materials

- 1 each as needed: red, blue, green, and yellow playing pieces
- 1 number cube
- Clue Cards

How to Play

1. Each player puts a playing piece on START.
2. To take a turn, toss the number cube. Move that many spaces.
3. If you land on these spaces:

   Green Space Follow the directions in the space.

   Yellow Space Explain how to evaluate the expression. If you are correct, move ahead 1 space.

   Blue Space Use a math term to name what is shown. If you are correct, move ahead 1 space.

   Red Space The player to your right draws a Clue Card and reads you the question. If you answer correctly, move ahead 1 space. Return the Clue Card to the bottom of the pile.

4. The first player to reach FINISH wins.

Word Box

algebraic
 expression
base
coefficient
difference
equivalent
 expressions
evaluate
exponent
factor
like terms
numerical
 expression
order of operations
product
quotient
sum
terms
variable

Game

START

DIRECTIONS Each player puts a playing piece on START. To take a turn, toss the number cube. Move that many spaces. • If you land on these spaces: Green Space – Follow the directions in the space. • Yellow Space – Explain how to evaluate the expression. If you are correct, move ahead 1 space. • Blue Space – Use a math term to name what is shown. If you are correct, move ahead 1 space. • Red Space – The player to your right draws a Clue Card and reads you the question. If you answer correctly, move ahead 1 space. Return the Clue Card to the bottom of the pile. • The first player to reach FINISH wins.

Take a scenic hike. Move ahead 1.

$4 \times (5 + 1)$

CLUE CARD

5^2

Visit the Natural Bridge. Go back 1.

$48 \div 4^2$

Get a flat tire. Lose 1 turn.

CLUE CARD

2^3

CLUE CARD

CLUE CARD

FINISH

$4 \times (y + 5)$
for $y = 11$

Climb
Chimney Rock.
Trade places
with another
player.

$x + 7$

CLUE
CARD

$2 \times (3^3 + 7)$

CLUE
CARD

$7^2 - (x^2 + 5)$
for $x = 2$

$4^2 \times 10 + 6$

Ride along
the Great Smoky
Mountains
Railroad. Take
another turn.

The Write Way

Reflect

Choose one idea. Write about it.

- Tell how to rewrite this expression so that each base has an exponent.

  $4 \times 4 \times 4 \times 7 \times 7$

- Explain how to evaluate this expression using the order of operations:

  $3 \times (1 + 5^2).$

- Define the term algebraic expression in your own words. Give an example.

- Jin wrote the following expression. Explain how he can simplify the expression by using like terms. Then write an equivalent expression.

  $2x + x + 9$

Name _____

Exponents

Essential Question How do you write and find the value of expressions involving exponents?

Common Core Expressions and Equations—6.EE.A.1
MATHEMATICAL PRACTICES
MP6, MP7, MP8

You can use an exponent and a base to show repeated multiplication of the same factor. An **exponent** is a number that tells how many times a number called the **base** is used as a repeated factor.

$$\underbrace{5 \times 5 \times 5}_{\text{3 repeated factors}} = 5\underset{\text{base}}{\overset{\text{exponent}}{3}}$$

Math Idea

- 5^2 can be read "the 2nd power of 5" or "5 squared."

- 5^3 can be read "the 3rd power of 5" or "5 cubed."

? Unlock the Problem Real World

The table shows the number of bonuses a player can receive in each level of a video game. Use an exponent to write the number of bonuses a player can receive in level D.

 Use an exponent to write 3 × 3 × 3 × 3.

The number _____ is used as a repeated factor.

3 is used as a factor _____ times.

Write the base and exponent. _____

So, a player can receive _____ bonuses in level D.

Level	Bonuses
A	3
B	3 × 3
C	3 × 3 × 3
D	3 × 3 × 3 × 3

Math Talk MATHEMATICAL PRACTICES 6

Explain How do you know which number to use as the base and which number to use as the exponent?

Try This! **Use one or more exponents to write the expression.**

A 7 × 7 × 7 × 7 × 7

The number _____ is used as a repeated factor.

7 is used as a factor _____ times.

Write the base and exponent. _____

B 6 × 6 × 8 × 8 × 8

The numbers _____ and _____ are used as repeated factors.

6 is used as a factor _____ times.

8 is used as a factor _____ times.

Write each base with its own exponent. 6☐ × 8☐

Example 1 Find the value.

A 10^3

STEP 1 Use repeated multiplication to write 10^3.

The repeated factor is _____. $10^3 = $ _____ × _____ × _____

Write the factor _____ times.

_____

STEP 2 Multiply.

Multiply each pair of factors, working from left to right.

$10 \times 10 \times 10 = $ _____ × 10

= _____

B 7^1

The repeated factor is _____. $7^1 = $ _____

Write the factor _____ time.

MATHEMATICAL PRACTICES 7

Look for a Pattern In 10^3, what do you notice about the value of the exponent and the product? Is there a similar pattern in other powers of 10? Explain.

 Example 2 Write 81 with an exponent by using 3 as the base.

STEP 1 Find the correct exponent.

Try 2. $3^2 = 3 \times 3 = $ _____

Try 3. $3^3 = $ _____ × _____ × _____ = _____

Try 4. $3^4 = $ _____ × _____ × _____ × _____ = _____

_____

STEP 2 Write using the base and exponent.

$81 = $ _____

---

1. Explain how to write repeated multiplication of a factor by using an exponent.

_____

_____

2. *THINK SMARTER* Is 5^2 equal to 2^5? Explain why or why not.

_____

3. *MATHEMATICAL PRACTICE 6* **Describe a Method** Describe how you could have solved the problem in Example 2 by using division.

_____

_____

Name _____

1. Write 2^4 by using repeated multiplication. Then find the value of 2^4.

$2^4 = 2 \times 2 \times$ _____ $\times$ _____ $=$ _____

Use one or more exponents to write the expression.

2. $7 \times 7 \times 7 \times 7$

3. $5 \times 5 \times 5 \times 5 \times 5$

4. $3 \times 3 \times 4 \times 4$

MATHEMATICAL PRACTICES ⑧

Generalize In 3^4, does it matter in what order you multiply the factors when finding the value? Explain.

On Your Own

Find the value.

5. 20^2

6. 82^1

7. 3^5

8. Write 32 as a number with an exponent by using 2 as the base.

Complete the statement with the correct exponent.

9. $5^{\square} = 125$

10. $16^{\square} = 16$

11. $30^{\square} = 900$

12. **MATHEMATICAL PRACTICE** ⑧ **Use Repeated Reasoning**
Find the values of 4^1, 4^2, 4^3, 4^4, and 4^5. Look for a pattern in your results and use it to predict the ones digit in the value of 4^6.

13. *THINK SMARTER* Select the expressions that are equivalent to 32. Mark all that apply.

Ⓐ 2^5

Ⓑ 8^4

Ⓒ $2^3 \times 4$

Ⓓ $2 \times 4 \times 4$

Connect to Science

Bacterial Growth

Bacteria are tiny, one-celled organisms that live almost everywhere on Earth. Although some bacteria cause disease, other bacteria are helpful to humans, other animals, and plants. For example, bacteria are needed to make yogurt and many types of cheese.

Under ideal conditions, a certain type of bacterium cell grows larger and then splits into 2 "daughter" cells. After 20 minutes, the daughter cells split, resulting in 4 cells. This splitting can happen again and again as long as conditions remain ideal.

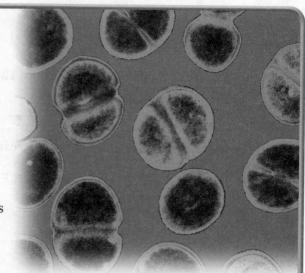

Complete the table.

Bacterial Growth	
Number of Cells	Time (min)
1	0
$2^1 = 2$	20
$2^2 = 2 \times 2 = 4$	40
$2^3 = \underline{} \times \underline{} \times \underline{} = \underline{}$	60
$2^{} = 2 \times 2 \times 2 \times 2 = 16$	80
$2^5 = \underline{} \times \underline{} \times \underline{} \times \underline{} = \underline{}$	100
$2^{} = \underline{} \times \underline{} \times \underline{} \times \underline{} \times \underline{} = \underline{}$	120
$2^7 = 2 \times 2 \times 2 \times 2 \times 2 \times 2 \times 2 = \underline{}$	\underline{}

Extend the pattern in the table above to answer 14 and 15.

14. **GO DEEPER** What power of 2 shows the number of cells after 3 hours? How many cells are there after 3 hours?

15. **THINK SMARTER** How many minutes would it take to have a total of 4,096 cells?

Math on the Spot

Exponents

Common Core
COMMON CORE STANDARD—6.EE.A.1
Apply and extend previous understandings of
arithmetic to algebraic expressions.

Use one or more exponents to write the expression.

1. 6×6

2. $11 \times 11 \times 11 \times 11$

3. $9 \times 9 \times 9 \times 9 \times 7 \times 7$

_____
6^2

Find the value.

4. 6^4

5. 1^6

6. 10^5

7. Write 144 with an exponent by using 12 as the base.

8. Write 343 with an exponent by using 7 as the base.

Problem Solving · Real World

9. Each day Sheila doubles the number of push-ups she did the day before. On the fifth day, she does $2 \times 2 \times 2 \times 2 \times 2$ push-ups. Use an exponent to write the number of push-ups Shelia does on the fifth day.

10. The city of Beijing has a population of more than 10^7 people. Write 10^7 without using an exponent.

11. **WRITE** ▸*Math* Explain what the expression 4^5 means and how to find its value.

Lesson Check (6.EE.A.1)

1. The number of games in the first round of a chess tournament is equal to $2 \times 2 \times 2 \times 2 \times 2 \times 2$. Write the number of games using an exponent.

2. The number of gallons of water in a tank at an aquarium is equal to 8^3. How many gallons of water are in the tank?

Spiral Review (6.RP.A.3a, 6.RP.A.3c, 6.RP.A.3d)

3. The table shows the amounts of strawberry juice and lemonade needed to make different amounts of strawberry lemonade. Name another ratio of strawberry juice to lemonade that is equivalent to the ratios in the table.

Strawberry juice (cups)	2	3	4
Lemonade (cups)	6	9	12

4. Which percent is equivalent to the fraction $\frac{37}{50}$?

5. How many milliliters are equivalent to 2.7 liters?

6. Use the formula $d = rt$ to find the distance traveled by a car driving at an average speed of 50 miles per hour for 4.5 hours.

FOR MORE PRACTICE
GO TO THE
Personal Math Trainer

Name _____

Evaluate Expressions Involving Exponents

Essential Question How do you use the order of operations to evaluate expressions involving exponents?

Common Core **Expressions and Equations—
6.EE.A.1**
MATHEMATICAL PRACTICES
MP4, MP6, MP7

A **numerical expression** is a mathematical phrase that uses only numbers and operation symbols.

$$3 + 16 \times 2^2 \qquad 4 \times (8 + 5^1) \qquad 2^3 + 4$$

You **evaluate** a numerical expression when you find its value. To evaluate an expression with more than one type of operation, you must follow a set of rules called the **order of operations**.

Order of Operations

1. Perform operations in parentheses.
2. Find the values of numbers with exponents.
3. Multiply and divide from left to right.
4. Add and subtract from left to right.

Unlock the Problem

An archer shoots 6 arrows at a target. Two arrows hit the ring worth 8 points, and 4 arrows hit the ring worth 4 points. Evaluate the expression $2 \times 8 + 4^2$ to find the archer's total number of points.

Follow the order of operations.

Write the expression. There are no parentheses.

$$2 \times 8 + 4^2$$

Find the value of numbers with exponents.

$$2 \times 8 + \_\_\_\_\_$$

_____ from left to right.

$$\_\_\_\_\_ + 16$$

Then add.

$$\_\_\_\_\_$$

So, the archer scores a total of _____ points.

Math Talk MATHEMATICAL PRACTICES ⑥

Explain In which order should you perform the operations to evaluate the expression $30 - 10 + 5^2$?

Try This! Evaluate the expression $24 \div 2^3$.

There are no parentheses.	$24 \div 2^3$
Find the value of numbers with exponents.	$24 \div \_\_\_\_\_$
Then divide.	$\_\_\_\_\_$

🔓 Example 1 — Evaluate the expression $72 \div (13 - 4) + 5 \times 2^3$.

Write the expression.

$$72 \div (13 - 4) + 5 \times 2^3$$

Perform operations in _____.

$$72 \div \text{_____} + 5 \times 2^3$$

Find the values of numbers with _____.

$$72 \div 9 + 5 \times \text{_____}$$

Multiply and _____ from left to right.

$$\text{_____} + 5 \times 8$$

$$8 + \text{_____}$$

Then add.

$$\text{_____}$$

🔓 Example 2

Last month, an online bookstore had approximately 10^5 visitors to its website. On average, each visitor bought 2 books. Approximately how many books did the bookstore sell last month?

STEP 1 Write an expression.

Think: The number of books sold is equal to the number of visitors times the number of books each visitor bought.

(number of visitors) (times) (number of books bought)

$$10^5 \qquad \times \qquad \text{_____}$$

STEP 2 Evaluate the expression.

Write the expression. There are no parentheses.

$$10^5 \times 2$$

Find the values of numbers with _____.

$$\text{_____} \times 2$$

Multiply.

$$\text{_____}$$

So, the bookstore sold approximately _____ books last month.

- **Explain** why the order of operations is necessary.

_____

_____

_____

© Houghton Mifflin Harcourt Publishing Company

Share and Show MATH BOARD

1. Evaluate the expression $9 + (5^2 - 10)$.

$9 + (5^2 - 10)$ Write the expression.

$9 + ($ _____ $- 10)$ Follow the order of operations within the parentheses.

$9 + $ _____

_____   Add.

Evaluate the expression.

2. $6 + 3^3 ÷ 9$

✓ **3.** $(15 - 3)^2 ÷ 9$

✓ **4.** $(8 + 9^2) - 4 × 10$

MATHEMATICAL PRACTICES 7

Look for Structure How does the parentheses make the values of these expressions different: $(2^2 + 8) ÷ 4$ and $2^2 + (8 ÷ 4)$?

On Your Own

Evaluate the expression.

5. $10 + 6^2 × 2 ÷ 9$

6. $6^2 - (2^3 + 5)$

7. $16 + 18 ÷ 9 + 3^4$

THINK SMARTER Place parentheses in the expression so that it equals the given value.

8. $10^2 - 50 ÷ 5$
value: 10

9. $20 + 2 × 5 + 4^1$
value: 38

10. $28 ÷ 2^2 + 3$
value: 4

Problem Solving • Applications

Use the table for 11–13.

11. 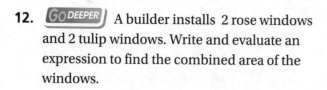 **Write an Expression** To find the cost of a window, multiply its area in square feet by the price per square foot. Write and evaluate an expression to find the cost of a knot window.

_____

12. **GO DEEPER** A builder installs 2 rose windows and 2 tulip windows. Write and evaluate an expression to find the combined area of the windows.

_____

13. **THINK SMARTER** DeShawn bought a tulip window. Emma bought a rose window. Write and evaluate an expression to determine how much more DeShawn paid for his window than Emma paid for hers.

_____

14. **What's the Error?** Darius wrote $17 - 2^2 = 225$. Explain his error.

_____

_____

_____

_____

15. **THINK SMARTER** Ms. Hall wrote the expression $2 \times (3 + 5)^2 \div 4$ on the board. Shyann said the first step is to evaluate 5^2. Explain Shyann's mistake. Then evaluate the expression.

_____

_____

 Math on the Spot

Art Glass Windows

Type	Area (square feet)	Price per square foot
Knot	2^2	$27
Rose	3^2	$30
Tulip	4^2	$33

 **WRITE** *Math* • **Show Your Work**

Evaluate Expressions Involving Exponents

COMMON CORE STANDARD—6.EE.A.1
Apply and extend previous understandings of arithmetic to algebraic expressions.

Evaluate the expression.

1. $5 + 17 - 10^2 \div 5$

$5 + 17 - 100 \div 5$

$5 + 17 - 20$

$22 - 20$

2

2. $7^2 - 3^2 \times 4$

3. $2^4 \div (7 - 5)$

4. $(8^2 + 36) \div (4 \times 5^2)$

5. $12 + 21 \div 3 + (2^2 \times 0)$

6. $(12 - 8)^3 - 24 \times 2$

Place parentheses in the expression so that it equals the given value.

7. $12 \times 2 + 2^3$; value: 120

8. $7^2 + 1 - 5 \times 3$; value: 135

Problem Solving Real World

9. Hugo is saving for a new baseball glove. He saves $10 the first week, and $6 each week for the next 6 weeks. The expression $10 + 6^2$ represents the total amount in dollars he has saved. What is the total amount Hugo has saved?

10. A scientist placed 5 fish eggs in a tank. Each day, twice the number of eggs from the previous day hatch. The expression 5×2^6 represents the number of eggs that hatch on the seventh day. How many eggs hatch on the seventh day?

11. **WRITE** ▶*Math* Explain how you could determine whether a calculator correctly performs the order of operations.

Lesson Check (6.EE.A.1)

1. Ritchie wants to paint his bedroom ceiling and four walls. The ceiling and each of the walls are 8 feet by 8 feet. A gallon of paint covers 40 square feet. Write an expression that can be used to find the number of gallons of paint Ritchie needs to buy.

_____

_____

2. A Chinese restaurant uses about 225 pairs of chopsticks each day. The manager wants to order a 30-day supply of chopsticks. The chopsticks come in boxes of 750 pairs. How many boxes should the manager order?

_____

_____

Spiral Review (6.RP.A.3a, 6.RP.A.3c, 6.RP.A.3d, 6.EE.A.1)

3. Annabelle spent $5 to buy 4 raffle tickets. How many tickets can she buy for $20?

_____

4. Gavin has 460 baseball players in his collection of baseball cards, and 15% of the players are pitchers. How many pitchers are in Gavin's collection?

_____

5. How many pounds are equivalent to 40 ounces?

_____

6. List the expressions in order from least to greatest.

$$1^5 \qquad 3^3 \qquad 4^2 \qquad 8^1$$

_____

FOR MORE PRACTICE
GO TO THE
Personal Math Trainer

Name _____

Write Algebraic Expressions

Essential Question How do you write an algebraic expression to represent a situation?

Expressions and Equations—6.EE.A.2a

MATHEMATICAL PRACTICES
MP2, MP6, MP7

An **algebraic expression** is a mathematical phrase that includes at least one variable. A **variable** is a letter or symbol that stands for one or more numbers.

$x + 10$ $3 \times y$ $3 \times (a + 4)$
↑   ↑   ↑
variable   variable   variable

> ## Math Idea
> There are several ways to show multiplication with a variable. Each expression below represents "3 times y."
>
> $3 \times y$ $3y$ $3(y)$ $3 \cdot y$

Unlock the Problem

An artist charges $5 for each person in a cartoon drawing. Write an algebraic expression for the cost in dollars for a drawing that includes p people.

🔑 **Write an algebraic expression for the cost.**

Think: | cost for each person | times | number of _____ |
           ↓   ↓   ↓
      _____   × p

So, the cost in dollars is _____.

Math Talk   MATHEMATICAL PRACTICES ②

Reasoning Why is p an appropriate variable for this problem? Would it be appropriate to select a different variable? Explain.

Try This! On Mondays, a bakery adds 2 extra muffins for free with every muffin order. Write an algebraic expression for the number of muffins customers will receive on Mondays when they order m muffins.

Think: | muffins ordered | _____ | extra muffins on Mondays |
         ↓   ↓   ↓
    _____   +   2

So, customers will receive _____ muffins on Mondays.

Example 1

The table at the right shows the number of points that items on a quiz are worth. Write an algebraic expression for the quiz score of a student who gets m multiple-choice items and s short-answer items correct.

Quiz Scoring	
Item Type	Points
Multiple-choice	2
Short-answer	5

points for multiple-choice items	_____	points for short-answer items
↓	↓	↓
$(2 \times m)$	+	(_____)

So, the student's quiz score is _____ points.

Example 2

Write an algebraic expression for the word expression.

A 30 more than the product of 4 and x

Think: Start with the product of 4 and x. Then find 30 more than the product.

the product of 4 and x _____ × _____

30 more than the product   _____ + 4x

B 4 times the sum of x and 30

Think: Start with the sum of x and 30. Then find 4 times the sum.

the sum of x and 30   _____ + _____

4 times the sum   _____ × (x + 30)

1. When you write an algebraic expression with two operations, how can you show which operation to do first?

_____

2. **THINK SMARTER** One student wrote $4 + x$ for the word expression "4 more than x." Another student wrote $x + 4$ for the same word expression. Are both students correct? Justify your answer.

_____

_____

_____

Name _____

1. Write an algebraic expression for the product of 6 and *p*.

   What operation does the word "product" indicate?

   _____

   The expression is _____ × _____.

Write an algebraic expression for the word expression.

✓ 2. 11 more than *e*

✓ 3. 9 less than the quotient of *n* and 5

_____

_____

On Your Own

Write an algebraic expression for the word expression.

> Math Talk
>
> MATHEMATICAL PRACTICES ⑥
>
> **Explain** why 3*x* is an algebraic expression.

4. 20 divided by *c*

5. 8 times the product of 5 and *t*

_____

_____

6. There are 12 eggs in a dozen. Write an algebraic expression for the number of eggs in *d* dozen.

7. A state park charges a $6.00 entry fee plus $7.50 per night of camping. Write an algebraic expression for the cost in dollars of entering the park and camping for *n* nights.

_____

_____

8. MATHEMATICAL PRACTICE ⑦ **Look for Structure** At a bookstore, the expression $2c + 8g$ gives the cost in dollars of *c* comic books and *g* graphic novels. Next month, the store's owner plans to increase the price of each graphic novel by $3. Write an expression that will give the cost of *c* comic books and *g* graphic novels next month.

_____

Unlock the Problem

9. Martina signs up for the cell phone plan described at the right. Write an expression that gives the total cost of the plan in dollars if Martina uses it for *m* months.

SPECIAL OFFER

CELL PHONE PLAN!

Pay a low monthly fee of **$50.**

Receive **$10** off your first month's fee.

a. What information do you know about the cell phone plan?

_____

_____

b. Write an expression for the monthly fee in dollars for *m* months.

_____

c. What operation can you use to show the discount of $10 for the first month?

_____

d. Write an expression for the total cost of the plan in dollars for *m* months.

_____

---

10. **THINK SMARTER** A group of *n* friends evenly share the cost of dinner. The dinner costs $74. After dinner, each friend pays $11 for a movie. Write an expression to represent what each friend paid for dinner and the movie.

_____

11. **THINK SMARTER** A cell phone company charges $40 per month plus $0.05 for each text message sent. Select the expressions that represent the cost in dollars for one month of cell phone usage and sending *m* text messages. Mark all that apply.

○ $40m + 0.05$

○ $40 + 0.05m$

○ 40 more than the product of 0.05 and *m*

○ the product of 40 and *m* plus 0.05

Write Algebraic Expressions

Common Core COMMON CORE STANDARD—6.EE.A.2a
Apply and extend previous understandings of arithmetic to algebraic expressions.

Write an algebraic expression for the word expression.

1. 13 less than p

   _____ $p - 13$ _____

2. the sum of x and 9

   _____

3. 6 more than the difference of b and 5

   _____

4. the sum of 15 and the product of 5 and v

   _____

5. the difference of 2 and the product of 3 and k

   _____

6. 12 divided by the sum of h and 2

   _____

7. the quotient of m and 7

   _____

8. 9 more than 2 multiplied by f

   _____

9. 6 minus the difference of x and 3

   _____

10. 10 less than the quotient of g and 3

    _____

11. the sum of 4 multiplied by a and 5 multiplied by b

    _____

12. 14 more than the difference of r and s

    _____

Problem Solving Real World

13. Let h represent Mark's height in inches. Suzanne is 7 inches shorter than Mark. Write an algebraic expression that represents Suzanne's height in inches.

    _____

14. A company rents bicycles for a fee of $10 plus $4 per hour of use. Write an algebraic expression for the total cost in dollars for renting a bicycle for h hours.

    _____

15. **WRITE** *Math* Give an example of a real-world situation involving two unknown quantities. Then write an algebraic expression to represent the situation.

    _____

    _____

Lesson Check (6.EE.A.2a)

1. The female lion at a zoo weighs 190 pounds more than the female cheetah. Let c represent the weight in pounds of the cheetah. Write an expression that gives the weight in pounds of the lion.

2. Tickets to a play cost $8 each. Write an expression that gives the ticket cost in dollars for a group of g girls and b boys.

Spiral Review (6.RP.A.2, 6.RP.A.3a, 6.RP.A.3c, 6.RP.A.3d, 6.EE.A.1)

3. A bottle of cranberry juice contains 32 fluid ounces and costs $2.56. What is the unit rate?

4. There are 32 peanuts in a bag. Elliott takes 25% of the peanuts from the bag. Then Zaire takes 50% of the remaining peanuts. How many peanuts are left in the bag?

5. Hank earns $12 per hour for babysitting. How much does he earn for 15 hours of babysitting?

6. Write an expression using exponents that represents the area of the figure in square centimeters.

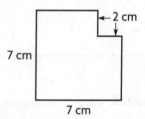

FOR MORE PRACTICE
GO TO THE
Personal Math Trainer

Name _____

Identify Parts of Expressions

Essential Question How can you describe the parts of an expression?

Common Core **Expressions and Equations—6.EE.A.2b**
MATHEMATICAL PRACTICES
MP2, MP6, MP7

Unlock the Problem

At a gardening store, seed packets cost $2 each. Martin bought 6 packets of lettuce seeds and 7 packets of pea seeds. The expression $2 \times (6 + 7)$ represents the cost in dollars of Martin's seeds. Identify the parts of the expression. Then write a word expression for $2 \times (6 + 7)$.

 Describe the parts of the expression $2 \times (6 + 7)$.

• Explain how you could find the cost of each type of seed.

_____

_____

_____

Identify the operations in the expression.

multiplication and _____

Describe the part of the expression in parentheses, and tell what it represents.

• The part in parentheses shows

the _____ of 6 and _____.

• The sum represents the number

of packets of _____

seeds plus the number of packets

of _____ seeds.

Describe the multiplication, and tell what it represents.

• One of the factors is _____. The other

factor is the _____ of 6 and 7.

• The product represents the _____ per packet times

the number of each type of _____ Martin bought.

So, a word expression for $2 \times (6 + 7)$ is "the _____ of 2 and the

_____ of _____ and 7."

• **MATHEMATICAL PRACTICE 6 Attend to Precision** Explain how the expression $2 \times (6 + 7)$ differs from $2 \times 6 + 7$. Then, write a word expression for $2 \times 6 + 7$.

_____

_____

_____

The **terms** of an expression are the parts of the expression that are separated by an addition or subtraction sign. A **coefficient** is a number that is multiplied by a variable.

4k + 5   The expression has two terms, 4k and 5. The coefficient of the term 4k is 4.

🔓 Example   Identify the parts of the expression. Then write a word expression for the algebraic expression.

A 2x + 8

Identify the terms in the expression.

The expression is the sum of _____ terms.

The terms are _____ and 8.

Describe the first term.

The first term is the product of the coefficient

_____ and the variable _____.

Describe the second term.

The second term is the number _____.

A word expression for 2x + 8 is "8 more than the _____

of _____ and x."

> **Math Talk**
>
> MATHEMATICAL PRACTICES ⑥
>
> **Explain** Why are the terms of the expression 2x and 8, not x and 8?

B 3a − 4b

Identify the terms in the expression.

The expression is the _____ of

2 terms. The terms are _____ and _____.

Describe the first term.

The first term is the product of the

_____ 3 and the variable _____.

Describe the second term.

The second term is the product of the

coefficient _____ and the variable _____.

A word expression for the algebraic expression is "the difference of

_____ times _____ and 4 _____ b."

> **Math Talk**
>
> MATHEMATICAL PRACTICES ②
>
> **Reasoning** Identify the coefficient of y in the expression 12 + y. Explain your reasoning.

Name _____

Identify the parts of the expression. Then, write a word expression for the numerical or algebraic expression.

1. $7 \times (9 \div 3)$

The part in parentheses shows the _____ of _____ and _____.

One factor of the multiplication is _____, and the other factor is $9 \div 3$.

Word expression: _____

2. $5m + 2n$

_____

_____

_____

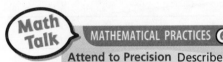

Math Talk

MATHEMATICAL PRACTICES ⑥

Attend to Precision Describe the expression $9 \times (a + b)$ as a product of two factors.

On Your Own

Practice: Copy and Solve Identify the parts of the expression. Then write a word expression for the numerical or algebraic expression.

3. $8 + (10 - 7)$ **4.** $1.5 \times 6 + 8.3$

5. $b + 12x$ **6.** $4a \div 6$

Identify the terms of the expression. Then, give the coefficient of each term.

7. $k - \frac{1}{3}d$ **8.** $0.5x + 2.5y$

_____   _____

_____   _____

9. MATHEMATICAL PRACTICE ② **Connect Symbols and Words** Ava said she wrote an expression with three terms. She said the first term has the coefficient 7, the second term has the coefficient 1, and the third term has the coefficient 0.1. Each term involves a different variable. Write an expression that could be the expression Ava wrote.

_____

Problem Solving • Applications

Use the table for 10–12.

10. GO DEEPER A football team scored 2 touchdowns and 2 extra points. Their opponent scored 1 touchdown and 2 field goals. Write a numerical expression for the points scored in the game.

_____

Football Scoring

Type	Points
Touchdown	6
Field Goal	3
Extra Point	1

11. Write an algebraic expression for the number of points scored by a football team that makes *t* touchdowns, *f* field goals, and *e* extra points.

_____

12. Identify the parts of the expression you wrote in Exercise 11.

_____

_____

_____

WRITE ▸ Math
Show Your Work

13. THINK SMARTER Give an example of an expression involving multiplication in which one of the factors is a sum. Explain why you do or do not need parentheses in your expression.

_____

_____

_____

14. THINK SMARTER Kennedy bought *a* pounds of almonds at $5 per pound and *p* pounds of peanuts at $2 per pound. Write an algebraic expression for the cost of Kennedy's purchase.

_____

Name _____

Identify Parts of Expressions

COMMON CORE STANDARD—6.EE.A.2a
Apply and extend previous understandings of arithmetic to algebraic expressions.

Identify the parts of the expression. Then write a word expression for the numerical or algebraic expression.

1. $(16 - 7) \div 3$

The subtraction is the difference of 16 and 7.

The division is the difference divided by 3.

Word expression: the difference of 16 and 7,

divided by 3

_____

_____

2. $8 + 6q + q$

_____

_____

_____

_____

_____

Identify the terms of the expression. Then give the coefficient of each term.

3. $11r + 7s$

_____

_____

4. $6g - h$

_____

_____

Problem Solving

5. Adam bought granola bars at the store. The expression $6p + 5n$ gives the number of bars in p boxes of plain granola bars and n boxes of granola bars with nuts. What are the terms of the expression?

_____

6. In the sixth grade, each student will get 4 new books. There is one class of 15 students and one class of 20 students. The expression $4 \times (15 + 20)$ gives the total number of new books. Write a word expression for the numerical expression.

_____

7. **WRITE** *Math* Explain how knowing the order of operations helps you write a word expression for a numerical or algebraic expression.

_____

_____

_____

Lesson Check (6.EE.A.2a)

1. A fabric store sells pieces of material for $5 each. Ali bought 2 white pieces and 8 blue pieces. She also bought a pack of buttons for $3. The expression $5 \times (2 + 8) + 3$ gives the cost in dollars of Ali's purchase. How can you describe the term $(2 + 8)$ in words?

2. A hotel offers two different types of rooms. The expression $k + 2f$ gives the number of beds in the hotel where k is the number of rooms with a king size bed and f is the number of rooms with 2 full size beds. What are the terms of the expression?

Spiral Review (6.RP.A.3b, 6.RP.A.3c, 6.RP.A.3d, 6.EE.A.2a)

3. Meg paid $9 for 2 tuna sandwiches. At the same rate, how much does Meg pay for 8 tuna sandwiches?

4. Jan is saving for a skateboard. She has saved $30 already, which is 20% of the total price. How much does the skateboard cost?

5. It took Eduardo 8 hours to drive from Buffalo, NY, to New York City, a distance of about 400 miles. Find his average speed.

6. Write an expression that represents the value, in cents, of n nickels.

**FOR MORE PRACTICE
GO TO THE
Personal Math Trainer**

Name _____

Evaluate Algebraic Expressions and Formulas

Essential Question How do you evaluate an algebraic expression or a formula?

Common Core Expressions and Equations—
6.EE.A.2c
MATHEMATICAL PRACTICES
MP1, MP6, MP8

To evaluate an algebraic expression, substitute numbers for the variables and then follow the order of operations.

Unlock the Problem (Real World)

Amir is saving money to buy an MP3 player that costs $120. He starts with $25, and each week he saves $9. The expression $25 + 9w$ gives the amount in dollars that Amir will have saved after w weeks.

- Which operations does the expression $25 + 9w$ include?

  _____

- In what order should you perform the operations?

  _____

A How much will Amir have saved after 8 weeks?

Evaluate the expression for $w = 8$.

Write the expression.	$25 + 9w$
Substitute 8 for w.	$25 + 9 \times$ _____
Multiply.	$25 +$ _____
Add.	_____

So, Amir will have saved $ _____ after 8 weeks.

B After how many weeks will Amir have saved enough money to buy the MP3 player?

Make a table to find the week when the amount saved is at least $120.

Week	Value of $25 + 9w$	Amount Saved
9	$25 + 9 \times 9 = 25 +$ _____ $= 106$	
10	$25 + 9 \times 10 = 25 +$ _____ $=$ _____	
11	$25 + 9 \times 11 = 25 +$ _____ $=$ _____	

So, Amir will have saved enough money for the

MP3 player after _____ weeks.

Math Talk

MATHEMATICAL PRACTICES 6

Explain What does it mean to substitute a value for a variable?

🔑 Example 1   Evaluate the expression for the given value of the variable.

Ⓐ $4 \times (m - 8) \div 3$ for $m = 14$

Write the expression.	$4 \times (m - 8) \div 3$
Substitute 14 for m.	$4 \times (\underline{\hspace{1cm}} - 8) \div 3$
Perform operations in parentheses.	$4 \times \underline{\hspace{1cm}} \div 3$
Multiply and divide from left to right.	$\underline{\hspace{1cm}} \div 3$
	$\underline{\hspace{1cm}}$

Ⓑ $3 \times (y^2 + 2)$ for $y = 4$

Write the expression.	$3 \times (y^2 + 2)$
Substitute 4 for y.	$3 \times (\underline{\hspace{1cm}}^2 + 2)$
Follow the order of operations within the parentheses.	$3 \times (\underline{\hspace{1cm}} + 2)$
	$3 \times \underline{\hspace{1cm}}$
Multiply.	$\underline{\hspace{1cm}}$

ERROR Alert

When squaring a number, be sure to multiply the number by itself.

$4^2 = 4 \times 4$

Recall that a *formula* is a set of symbols that expresses a mathematical rule.

🔑 Example 2

The formula $P = 2\ell + 2w$ gives the perimeter P of a rectangle with length ℓ and width w. What is the perimeter of a rectangular garden with a length of 2.4 meters and a width of 1.2 meters?

Write the expression for the perimeter of a rectangle.	$2\ell + 2w$
Substitute 2.4 for ℓ and $\underline{\hspace{1cm}}$ for w.	$2 \times \underline{\hspace{1cm}} + 2 \times \underline{\hspace{1cm}}$
Multiply from left to right.	$\underline{\hspace{1cm}} + 2 \times 1.2$
	$4.8 + \underline{\hspace{1cm}}$
Add.	$\underline{\hspace{1cm}}$

So, the perimeter of the garden is $\underline{\hspace{1cm}}$ meters.

Math Talk

MATHEMATICAL PRACTICES ⑥

Compare How is evaluating an algebraic expression different from evaluating a numerical expression?

Name _____

1. Evaluate $5k + 6$ for $k = 4$.

Write the expression.   _____

Substitute 4 for k. $5 \times$ _____ $+ 6$

Multiply.   _____ $+ 6$

Add.   _____

Evaluate the expression for the given value of the variable.

2. $m - 9$ for $m = 13$

3. $16 - 3b$ for $b = 4$

4. $p^2 + 4$ for $p = 6$

5. The formula $A = \ell w$ gives the area A of a rectangle with length ℓ and width w. What is the area in square feet of a United States flag with a length of 12 feet and a width of 8 feet?

Math Talk

MATHEMATICAL PRACTICES ⑧

Use Repeated Reasoning
What information do you need to evaluate an algebraic expression?

On Your Own

Practice: Copy and Solve Evaluate the expression for the given value of the variable.

6. $7s + 5$ for $s = 3$

7. $21 - 4d$ for $d = 5$

8. $(t - 6)^2$ for $t = 11$

9. $6 \times (2v - 3)$ for $v = 5$

10. $2 \times (k^2 - 2)$ for $k = 6$

11. $5 \times (f - 32) \div 9$ for $f = 95$

12. GO DEEPER The formula $P = 4s$ gives the perimeter P of a square with side length s. How much greater is the perimeter of a square with a side length of $5\frac{1}{2}$ inches than a square with a side length of 5 inches?

© Houghton Mifflin Harcourt Publishing Company

Problem Solving • Applications

The table shows how much a company charges for skateboard wheels. Each pack of 8 wheels costs $50. Shipping costs $7 for any order. Use the table for 13–15.

13. Complete the table.

14. A skateboard club has $200 to spend on new wheels this year. What is the greatest number of packs of wheels the club can order?

_____

15.  **Make Sense of Problems** A sporting goods store placed an order for 12 packs of wheels on the first day of each month last year. How much did the sporting goods store spend on these orders last year?

_____

Costs for Skateboard Wheels		
Packs	$50 \times n + 7$	Cost
1	$50 \times 1 + 7$	$57
2		
3		
4		
5		

16. THINK SMARTER **What's the Error?** Bob used these steps to evaluate $3m - 3 \div 3$ for $m = 8$. Explain his error.

$3 \times 8 - 3 \div 3 = 24 - 3 \div 3$

$= 21 \div 3$

$= 7$

Math on the Spot

WRITE Math • Show Your Work

_____

_____

_____

_____

17. THINK SMARTER The surface area of a cube can be found by using the formula $6s^2$, where s represents the length of the side of the cube.

The surface area of a cube that has a side length of

3 meters is

54
108
2,916

meters squared.

Evaluate Algebraic Expressions and Formulas

Common Core **COMMON CORE STANDARD—6.EE.A.2c**
Apply and extend previous understandings of arithmetic to algebraic expressions.

Evaluate the expression for the given values of the variables.

1. $w + 6$ for $w = 11$

$11 + 6$

17

2. $17 - 2c$ for $c = 7$

_____

3. $b^2 - 4$ for $b = 5$

_____

4. $(h - 3)^2$ for $h = 5$

_____

5. $m + 2m + 3$ for $m = 12$

_____

6. $4 \times (21 - 3h)$ for $h = 5$

_____

7. $7m - 9n$ for $m = 7$ and $n = 5$

_____

8. $d^2 - 9k + 3$ for $d = 10$ and $k = 9$

_____

9. $3x + 4y \div 2$ for $x = 7$ and $y = 10$

_____

Problem Solving · Real World

10. The formula $P = 2\ell + 2w$ gives the perimeter P of a rectangular room with length ℓ and width w. A rectangular living room is 26 feet long and 21 feet wide. What is the perimeter of the room?

_____

11. The formula $C = 5(F - 32) \div 9$ gives the Celsius temperature in C degrees for a Fahrenheit temperature of F degrees. What is the Celsius temperature for a Fahrenheit temperature of 122 degrees?

_____

12. **WRITE** ▸Math Explain how the terms *variable, algebraic expression,* and *evaluate* are related.

_____

_____

_____

Lesson Check (6.EE.A.2c)

1. When Debbie baby-sits, she charges $5 to go to the house plus $8 for every hour she is there. The expression $5 + 8h$ gives the amount in dollars she charges. How much will she charge to baby-sit for 5 hours?

2. The formula to find the cost C in dollars of a square sheet of glass is $C = 25s^2$ where s represents the length of a side in feet. How much will Ricardo pay for a square sheet of glass that is 3 feet on each side?

Spiral Review (6.NS.A.1, 6. RP.A.3c, 6.EE.A.1, 6.EE.A.2a)

3. Evaluate using the order of operations.

$$\frac{3}{4} + \frac{5}{6} \div \frac{2}{3}$$

4. Patricia scored 80% on a math test. She missed 4 problems. How many problems were on the test?

5. What is the value of 7^3?

6. James and his friends ordered b hamburgers that cost $4 each and f fruit cups that cost $3 each. Write an algebraic expression for the total cost in dollars of their purchases.

FOR MORE PRACTICE
GO TO THE
Personal Math Trainer

Name _____

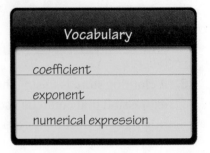

Vocabulary

Vocabulary
coefficient
exponent
numerical expression

Choose the best term from the box to complete the sentence.

1. A(n) _____ tells how many times a base is used as a factor. (p. 357)

2. The mathematical phrase $5 + 2 \times 18$ is an example of a(n)

   _____ . (p. 363)

Concepts and Skills

Find the value. (6.EE.A.1)

3. 5^4

4. 21^2

5. 8^3

_____ _____ _____

Evaluate the expression. (6.EE.A.1)

6. $9^2 \times 2 - 4^2$

7. $2 \times (10 - 2) \div 2^2$

8. $30 - (3^3 - 8)$

_____ _____ _____

Write an algebraic expression for the word expression. (6.EE.A.2a)

9. the quotient of c and 8

10. 16 more than the product of 5 and p

11. 9 less than the sum of x and 5

_____ _____ _____

Evaluate the expression for the given value of the variable. (6.EE.A.2c)

12. $5 \times (h + 3)$ for $h = 7$

13. $2 \times (c^2 - 5)$ for $c = 4$

14. $7a - 4a$ for $a = 8$

_____ _____ _____

15. The greatest value of any U.S. paper money ever printed is 10^5 dollars. What is this amount written in standard form? (6.EE.A.1)

---

16. A clothing store is raising the price of all its sweaters by $3.00. Write an expression that could be used to find the new price of a sweater that originally cost d dollars. (6.EE.A.2a)

---

17. Kendra bought a magazine for $3 and 4 paperback books for $5 each. The expression $3 + 4 \times 5$ represents the total cost in dollars of her purchases. What are the terms in this expression? (6.EE.A.2b)

---

18. The expression $5c + 7m$ gives the number of people who can ride in c cars and m minivans. What are the coefficients in this expression? (6.EE.A.2b)

---

19. GO DEEPER The formula $P = a + b + c$ gives the perimeter P of a triangle with side lengths a, b, and c. How much greater is the perimeter of a triangular field with sides that measure 33 yards, 56 yards, and 65 yards than the perimeter of a triangular field with sides that measure 26 yards, 49 yards, and 38 yards? (6.EE.A.2c)

Use Algebraic Expressions

Essential Question How can you use variables and algebraic expressions to solve problems?

Common Core Expressions and Equations—
6.EE.B.6
MATHEMATICAL PRACTICES
MP1, MP2, MP6

Sometimes you have an unknown number that you need to solve a problem. You can represent a problem like this by writing an algebraic expression in which a variable represents the unknown number.

🔑 Unlock the Problem

Rafe's flight from Los Angeles to New York took 5 hours. He wants to know the average speed of the plane in miles per hour.

A **Write an expression to represent the average speed of the plane.**

🔑 **Use a variable to represent the unknown quantity.**

Think: The plane's average speed is equal to the distance traveled divided by the time traveled.

Use a variable to represent the unknown quantity.

Let d represent the _____

traveled in units of _____.

Write an algebraic expression for the average speed.

$$\frac{d \text{ mi}}{\boxed{} \text{ hr}}$$

B **Rafe looks up the distance between Los Angeles and New York on the Internet and finds that the distance is 2,460 miles. Use this distance to find the average speed of Rafe's plane.**

🔑 **Evaluate the expression for $d = 2{,}460$.**

Write the expression.

$$\frac{d \text{ mi}}{5 \text{ hr}}$$

Substitute 2,460 for d.

$$\frac{\boxed{} \text{ mi}}{5 \text{ hr}}$$

Divide to find the unit rate.

$$\frac{2{,}460 \text{ mi} \div \boxed{}}{5 \text{ hr} \div 5} = \frac{\boxed{} \text{ mi}}{1 \text{ hr}}$$

So, the plane's average speed was _____ miles per hour.

Math Talk MATHEMATICAL PRACTICES ①

Evaluate How could you check whether you found the plane's average speed correctly?

In the problem on the previous page, the variable represented a single value—the distance in miles between Los Angeles and New York. In other situations, a variable may represent any number in a particular set of numbers, such as the set of positive numbers.

Example Joanna makes and sells candles online. She charges $7 per candle, and shipping is $5 per order.

A Write an expression that Joanna can use to find the total cost for any candle order.

Think: The number of candles a customer buys will vary from order to order.

Let n represent the number of _____ a customer buys, where n is a whole number greater than 0.

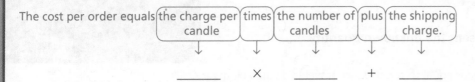

The cost per order equals | the charge per candle | times | the number of candles | plus | the shipping charge.

_____ × _____ + _____

So, an expression for the total cost of a candle order is _____.

B In March, one of Joanna's customers placed an order for 4 candles. In May, the same customer placed an order for 6 candles. What was the total charge for both orders?

STEP 1 Find the charge in dollars for each order.

	March	May
Write the expression.	$7n + 5$	$7n + 5$
Substitute the number of candles ordered for n.	$7 \times$ _____ $+ 5$	$7 \times$ _____ $+ 5$
Follow the order of operations.	_____ $+ 5$	_____ $+ 5$
	_____	_____

STEP 2 Find the charge in dollars for both orders.

Add the charge in dollars for March to the charge in dollars for May.

_____ + _____ = _____

So, the total charge for both orders was _____.

Math Talk

MATHEMATICAL PRACTICES ②

Reasoning Why is the value of the variable n in the Example restricted to the set of whole numbers greater than 0?

Name _____

Louisa read that the highest elevation of Mount Everest is 8,848 meters. She wants to know how much higher Mount Everest is than Mount Rainier. Use this information for 1–2.

1. Write an expression to represent the difference in heights of the two mountains. Tell what the variable in your expression represents.

_____

_____

2. Louisa researches the highest elevation of Mount Rainier and finds that it is 4,392 meters. Use your expression to find the difference in the mountains' heights.

_____

Math Talk   MATHEMATICAL PRACTICES ②

Reason Quantitatively Explain whether the variable in Exercise 1 represents a single unknown number or any number in a particular set.

On Your Own

A muffin recipe calls for 3 times as much flour as sugar. Use this information for 3–5.

3. Write an expression that can be used to find the amount of flour needed for a given amount of sugar. Tell what the variable in your expression represents.

_____

_____

4. Use your expression to find the amount of flour needed when $\frac{3}{4}$ cup of sugar is used.

_____

5. MATHEMATICAL PRACTICE ② **Reason Quantitatively** Is the value of the variable in your expression restricted to a particular set of numbers? Explain.

_____

_____

Practice: Copy and Solve Write an algebraic expression for each word expression. Then evaluate the expression for these values of the variable: $\frac{1}{2}$, 4, and 6.5.

6. the quotient of p and 4

7. 4 less than the sum of x and 5

Problem Solving • Applications (Real World)

Use the graph for 8–10.

8. Write expressions for the distance in feet that each animal could run at top speed in a given amount of time. Tell what the variable in your expressions represents.

_____

_____

_____

_____

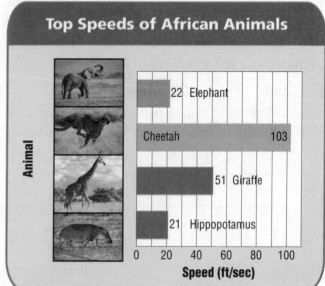

Top Speeds of African Animals

Animal / Speed (ft/sec)

- Elephant: 22
- Cheetah: 103
- Giraffe: 51
- Hippopotamus: 21

9. **GO DEEPER** How much farther could a cheetah run in 20 seconds at top speed than a hippopotamus could?

_____

10. **THINK SMARTER** A giraffe runs at top speed toward a tree that is 400 feet away. Write an expression that represents the giraffe's distance in feet from the tree after s seconds.

_____

| WRITE ▸ Math • Show Your Work |

Personal Math Trainer

11. **THINK SMARTER +** A carnival charges $7 for admission and $2 for each ride. An expression for the total cost of going to the carnival and riding n rides is $7 + 2n$.

Complete the table by finding the total cost of going to the carnival and riding n rides.

Number of rides, n	$7 + 2n$	Total Cost
1		
2		
3		
4		

Use Algebraic Expressions

Common Core

COMMON CORE STANDARD—6.EE.B.6
*Reason about and solve one-variable
equations and inequalities.*

Jeff sold the pumpkins he grew for $7 each at the farmer's market.

1. Write an expression to represent the amount of money in dollars Jeff made selling the pumpkins. Tell what the variable in your expression represents.

   7p, where p is the number

   of pumpkins

2. If Jeff sold 30 pumpkins, how much money did he make?

_____

An architect is designing a building. Each floor will be 12 feet tall.

3. Write an expression for the number of floors the building can have for a given building height. Tell what the variable in your expression represents.

_____

4. If the architect is designing a building that is 132 feet tall, how many floors can be built?

_____

Write an algebraic expression for each word expression. Then evaluate the expression for these values of the variable: 1, 6, 13.5.

5. the quotient of 300 and the sum of b and 24

_____

6. 13 more than the product of m and 5

_____

Problem Solving · Real World

7. In the town of Pleasant Hill, there is an average of 16 sunny days each month. Write an expression to represent the approximate number of sunny days for any number of months. Tell what the variable represents.

_____

8. How many sunny days can a resident of Pleasant Hill expect to have in 9 months?

_____

9. **WRITE** ▸*Math* Describe a situation in which a variable could be used to represent any whole number greater than 0.

_____

_____

Lesson Check (6.EE.B.6)

1. Oliver drives 45 miles per hour. Write an expression that represents the distance in miles he will travel for h hours driven.

_____

_____

_____

2. Socks cost \$5 per pair. The expression $5p$ represents the cost in dollars of p pairs of socks. Why must p be a whole number?

_____

_____

_____

Spiral Review (6.RP.A.3c, 6.RP.A.3d, 6.EE.A.1, 6.EE.A.2c)

3. Sterling silver consists of 92.5% silver and 7.5% copper. What decimal represents the portion of silver in sterling silver?

_____

4. How many pints are equivalent to 3 gallons?

_____

5. Which operation should be done first to evaluate $10 + (66 - 6^2)$?

_____

6. Evaluate the algebraic expression $h(m + n) \div 2$ for $h = 4$, $m = 5$, and $n = 6$.

_____

**FOR MORE PRACTICE
GO TO THE
Personal Math Trainer**

Problem Solving • Combine Like Terms

Essential Question How can you use the strategy *use a model* to combine like terms?

Common Core

Expressions and Equations—
6.EE.A.3

MATHEMATICAL PRACTICES
MP3, MP4, MP7

Like terms are terms that have the same variables with the same exponents. Numerical terms are also like terms.

Algebraic Expression	Terms	Like Terms
$5x + 3y - 2x$	$5x$, $3y$, and $2x$	$5x$ and $2x$
$8z^2 + 4z + 12z^2$	$8z^2$, $4z$, and $12z^2$	$8z^2$ and $12z^2$
$15 - 3x + 5$	15, $3x$, and 5	15 and 5

Unlock the Problem — Real World

Baseball caps cost $9, and patches cost $4. Shipping is $8 per order. The expression $9n + 4n + 8$ gives the cost in dollars of buying caps with patches for n players. Simplify the expression $9n + 4n + 8$ by combining like terms.

Use the graphic organizer to help you solve the problem.

Read the Problem

What do I need to find?	**What information do I need to use?**	**How will I use the information?**
I need to simplify the expression _____ .	I need to use the like terms $9n$ and _____ .	I can use a bar model to find the sum of the _____ terms.

Solve the Problem

Draw a bar model to add _____ and _____ . Each square represents n, or $1n$.

9n									4n			
n	n	n	n	n	n	n	n	n	n	n	n	n

_____n

The model shows that $9n + 4n =$ _____ . $9n + 4n + 8 =$ _____ $+ 8$

Math Talk

MATHEMATICAL PRACTICES ④

Use Models Explain how the bar model shows that your answer is correct.

So, a simplified expression for the cost in dollars is _____ .

🔓 Try Another Problem

Paintbrushes normally cost $5 each, but they are on sale for $1 off.
A paintbrush case costs $12. The expression $5p - p + 12$ can be used to find
the cost in dollars of buying p paintbrushes on sale plus a case for them.
Simplify the expression $5p - p + 12$ by combining like terms.

Use the graphic organizer to help you solve the problem.

Read the Problem

What do I need to find?	What information do I need to use?	How will I use the information?

Solve the Problem

So, a simplified expression for the cost in dollars is _____ .

1. **MATHEMATICAL PRACTICE ④** **Use Models** Explain how the bar model shows that your answer is correct.

_____

_____

_____

2. Explain how you could combine like terms without using a model.

_____

_____

_____

Name _____

Unlock the Problem

✓ Read the entire problem carefully before you begin to solve it.

✓ Check your answer by using a different method.

Share and Show 📋 MATH BOARD

1. Museum admission costs $7, and tickets to the mammoth exhibit cost $5. The expression $7p + 5p$ represents the cost in dollars for p people to visit the museum and attend the exhibit. Simplify the expression by combining like terms.

   First, draw a bar model to combine the like terms.

   Next, use the bar model to simplify the expression.

   _____

   So, a simplified expression for the cost in dollars is _____.

WRITE ▸ *Math*
Show Your Work

2. [THINK SMARTER] **What if** the cost of tickets to the exhibit were reduced to $3? Write an expression for the new cost in dollars for p people to visit the museum and attend the exhibit. Then, simplify the expression by combining like terms.

   _____

3. A store receives tomatoes in boxes of 40 tomatoes each. About 4 tomatoes per box cannot be sold due to damage. The expression $40b - 4b$ gives the number of tomatoes that the store can sell from a shipment of b boxes. Simplify the expression by combining like terms.

   _____

4. Each cheerleading uniform includes a shirt and a skirt. Shirts cost $12 each, and skirts cost $18 each. The expression $12u + 18u$ represents the cost in dollars of buying u uniforms. Simplify the expression by combining like terms.

   _____

5. A shop sells vases holding 9 red roses and 6 white roses. The expression $9v + 6v$ represents the total number of roses needed for v vases. Simplify the expression by combining like terms.

   _____

On Your Own

6. GO DEEPER Marco received a gift card. He used it to buy 2 bike lights for $10.50 each. Then he bought a handlebar bag for $18.25. After these purchases, he had $0.75 left on the card. How much money was on the gift card when Marco received it?

_____

7. Lydia collects shells. She has 24 sea snail shells, 16 conch shells, and 32 scallop shells. She wants to display the shells in equal rows, with only one type of shell in each row. What is the greatest number of shells Lydia can put in each row?

_____

Sea snail shells

Scallop shell

8. THINK SMARTER The three sides of a triangle measure $3x + 6$ inches, $5x$ inches, and $6x$ inches. Write an expression for the perimeter of the triangle in inches. Then simplify the expression by combining like terms.

_____

Conch shell

9. MATHEMATICAL PRACTICE ③ **Verify the Reasoning of Others** Karina states that you can simplify the expression $20x + 4$ by combining like terms to get $24x$. Does Karina's statement make sense? Explain.

_____

_____

_____

Personal Math Trainer

10. THINK SMARTER + Vincent is ordering accessories for his surfboard. A set of fins costs $24 each and a leash costs $15. The shipping cost is $4 per order. The expression $24b + 15b + 4$ can be used to find the cost in dollars of buying b fins and b leashes plus the cost of shipping.

For numbers 10a–10c, select True or False for each statement.

10a.   The terms are $24b$, $15b$ and 4.   ○ True   ○ False

10b.   The like terms are $24b$ and $15b$.   ○ True   ○ False

10c.   The simplified expression is $43b$.   ○ True   ○ False

Problem Solving • Combine Like Terms

COMMON CORE STANDARD—6.EE.A.3
Apply and extend previous understandings of
arithmetic to algebraic expressions.

Read each problem and solve.

1. A box of pens costs $3 and a box of markers costs $5. The
   expression $3p + 5p$ represents the cost in dollars to make p
   packages that includes 1 box of pens and 1 box of markers.
   Simplify the expression by combining like terms.

   $$3p + 5p = 8p$$

2. Riley's parents got a cell phone plan that has a $40 monthly
   fee for the first phone. For each extra phone, there is a $15
   phone service charge and a $10 text service charge. The
   expression $40 + 15e + 10e$ represents the total phone bill in
   dollars, where e is the number of extra phones. Simplify the
   expression by combining like terms.

   _____

3. A radio show lasts for h hours. For every 60 minutes of air
   time during the show, there are 8 minutes of commercials.
   The expression $60h - 8h$ represents the air time in minutes
   available for talk and music. Simplify the expression by
   combining like terms.

   _____

4. A sub shop sells a meal that includes an Italian sub for
   $6 and chips for $2. If a customer purchases more than
   3 meals, he or she receives a $5 discount. The expression $6m
   + 2m - 5$ shows the cost in dollars of the customer's order for
   m meals, where m is greater than 3. Simplify the expression by
   combining like terms.

   _____

5. **WRITE** ▸*Math* Explain how combining like terms is similar
   to adding and subtracting whole numbers. How are they different?

   _____

   _____

   _____

Lesson Check

1. For each gym class, a school has 10 soccer balls and 6 volleyballs. All of the classes share 15 basketballs. The expression $10c + 6c + 15$ represents the total number of balls the school has for c classes. What is a simpler form of the expression?

2. A public library wants to place 4 magazines and 9 books on each display shelf. The expression $4s + 9s$ represents the total number of items that will be displayed on s shelves. Simplify this expression.

Spiral Review

3. A bag has 8 bagels. Three of the bagels are cranberry. What percent of the bagels are cranberry?

4. How many kilograms are equivalent to 3,200 grams?

5. Toni earns $200 per week plus $5 for every magazine subscription that she sells. Write an expression that represents how much she will earn in dollars in a week in which she sells s subscriptions.

6. At a snack stand, drinks cost $1.50. Write an expression that could be used to find the total cost in dollars of d drinks.

FOR MORE PRACTICE
GO TO THE
Personal Math Trainer

Generate Equivalent Expressions

Essential Question How can you use properties of operations to write equivalent algebraic expressions?

Common Core **Expressions and Equations—**
6.EE.A.3
MATHEMATICAL PRACTICES
MP4, MP6, MP8

Equivalent expressions are equal to each other for any values of their variables. For example, $x + 3$ and $3 + x$ are equivalent. You can use properties of operations to write equivalent expressions.

$x + 3$	$3 + x$
$4 + 3$	$3 + 4$
7	7

Properties of Addition

Commutative Property of Addition If the order of terms changes, the sum stays the same.	$12 + a = a + 12$
Associative Property of Addition When the grouping of terms changes, the sum stays the same.	$5 + (8 + b) = (5 + 8) + b$
Identity Property of Addition The sum of 0 and any number is that number.	$0 + c = c$

Properties of Multiplication

Commutative Property of Multiplication If the order of factors changes, the product stays the same.	$d \times 9 = 9 \times d$
Associative Property of Multiplication When the grouping of factors changes, the product stays the same.	$11 \times (3 \times e) = (11 \times 3) \times e$
Identity Property of Multiplication The product of 1 and any number is that number.	$1 \times f = f$

! Unlock the Problem (Real World)

Nelson ran 2 miles, 3 laps, and 5 miles. The expression $2 + 3\ell + 5$ represents the total distance in miles Nelson ran, where ℓ is the length in miles of one lap. Write an equivalent expression with only two terms.

 Rewrite the expression $2 + 3\ell + 5$ with only two terms.

The like terms are 2 and _____. Use the

_____ Property to reorder the terms.

$$2 + 3\ell + 5 = 3\ell + \text{_____} + 5$$

Use the _____ Property to regroup the terms.

$$= 3\ell + (\text{\_\_\_\_\_} + \text{\_\_\_\_\_})$$

Add within the parentheses.

$$= 3\ell + \text{_____}$$

So, an equivalent expression for the total distance in miles is _____.

Distributive Property

Multiplying a sum by a number is the same as multiplying each term by the number and then adding the products.

$5 \times (g + 9) = (5 \times g) + (5 \times 9)$

The Distributive Property can also be used with multiplication and subtraction. For example, $2 \times (10 - h) = (2 \times 10) - (2 \times h)$.

🔑 Example 1 Use properties of operations to write an expression equivalent to $5a + 8a - 16$ by combining like terms.

Use the Commutative Property of Multiplication to rewrite the like terms $5a$ and $8a$.

$5a + 8a - 16 = a \times \underline{\hspace{1cm}} + a \times \underline{\hspace{1cm}} - 16$

Use the Distributive Property to rewrite $a \times 5 + a \times 8$.

$= \underline{\hspace{1cm}} \times (5 + 8) - 16$

Add within the parentheses.

$= a \times \underline{\hspace{1cm}} - 16$

Use the Commutative Property of Multiplication to rewrite $a \times 13$.

$= \underline{\hspace{1cm}} - 16$

So, the expression _____ is equivalent to $5a + 8a - 16$.

🔑 Example 2 Use the Distributive Property to write an equivalent expression.

Ⓐ $6(y + 7)$

Use the Distributive Property.

$6(y + 7) = (6 \times \underline{\hspace{1cm}}) + (6 \times \underline{\hspace{1cm}})$

Multiply within the parentheses.

$= 6y + \underline{\hspace{1cm}}$

So, the expression _____ is equivalent to $6(y + 7)$.

Math Idea

When one factor in a product is in parentheses, you can leave out the multiplication sign. So, $6 \times (y + 7)$ can be written as $6(y + 7)$.

Ⓑ $12a + 8b$

Find the greatest common factor (GCF) of the coefficients of the terms.

The GCF of 12 and 8 is _____.

Write the first term, $12a$, as the product of the GCF and another factor.

$12a + 8b = 4 \times 3a + 8b$

Write the second term, $8b$, as the product of the GCF and another factor.

$= 4 \times 3a + 4 \times \underline{\hspace{1cm}}$

Use the Distributive Property.

$= 4 \times (\underline{\hspace{1cm}} + 2b)$

So, the expression _____ is equivalent to $12a + 8b$.

Math Talk MATHEMATICAL PRACTICES ⑧

Generalize Give a different expression that is equivalent to $12a + 8b$. Explain what property you used.

Name _____

Use properties of operations to write an equivalent expression by combining like terms.

1. $3\frac{7}{10}r - 1\frac{5}{10}r$

_____

✓2. $20a + 18 + 16a$

_____

3. $7s + 8t + 10s + 12t$

_____

Use the Distributive Property to write an equivalent expression.

✓ 4. $8(h + 1.5)$

_____

5. $4m + 4p$

_____

6. $3a + 9b$

_____

On Your Own

Math Talk MATHEMATICAL PRACTICES ⑥

Compare List three expressions with two terms that are equivalent to 5x. Compare and discuss your list with a partner's.

Practice: Copy and Solve Use the Distributive Property to write an equivalent expression.

7. $3.5(w + 7)$

8. $\frac{1}{2}(f + 10)$

9. $4(3z + 2)$

10. $20b + 16c$

11. $30d + 18$

12. $24g - 8h$

13. MATHEMATICAL PRACTICE ④ **Write an Expression** The lengths of the sides of a triangle are $3t$, $2t + 1$, and $t + 4$. Write an expression for the perimeter (sum of the lengths). Then, write an equivalent expression with 2 terms.

_____

14. GO DEEPER Use properties of operations to write an expression equivalent to the sum of the expressions $3(g + 5)$ and $2(3g - 6)$.

_____

Problem Solving • Applications (Real World)

15. **THINK SMARTER** **Sense or Nonsense** Peter and Jade are using what they know about properties to write an expression equivalent to $2 \times (n + 6) + 3$. Whose answer makes sense? Whose answer is nonsense? **Explain** your reasoning.

Peter's Work

Expression:	$2 \times (n + 6) + 3$
Associative Property of Addition:	$2 \times n + (6 + 3)$
Add within parentheses:	$2 \times n + 9$
Multiply:	$2n + 9$

Jade's Work

Expression:	$2 \times (n + 6) + 3$
Distributive Property:	$(2 \times n) + (2 \times 6) + 3$
Multiply within parentheses:	$2n + 12 + 3$
Associative Property of Addition:	$2n + (12 + 3)$
Add within parentheses:	$2n + 15$

For the answer that is nonsense, correct the statement.

_____

_____

16. **THINK SMARTER** Write the algebraic expression in the box that shows an equivalent expression.

$6(z + 5)$	$6z + 5z$	$2 + 6z + 3$
$6z + 5$	$11z$	$6z + 30$

Generate Equivalent Expressions

Common Core

COMMON CORE STANDARD—6.EE.A.3
*Apply and extend previous understandings of
arithmetic to algebraic expressions.*

Use properties of operations to write an equivalent expression by combining like terms.

1. $7h - 3h$ **2.** $5x + 7 + 2x$ **3.** $16 + 13p - 9p$

_____ $4h$ _____   _____   _____

4. $y^2 + 13y - 8y$ **5.** $5(2h + 3) + 3h$ **6.** $12 + 18n + 7 - 14n$

_____   _____   _____

Use the Distributive Property to write an equivalent expression.

7. $2(9 + 5k)$ **8.** $4d + 8$ **9.** $21p + 35q$

_____   _____   _____

Problem Solving (Real World)

10. The expression $15n + 12n + 100$ represents the total cost in dollars for skis, boots, and a lesson for n skiers. Simplify the expression $15n + 12n + 100$. Then find the total cost for 8 skiers.

11. Casey has n nickels. Megan has 4 times as many nickels as Casey has. Write an expression for the total number of nickels Casey and Megan have. Then simplify the expression.

_____   _____

12. **WRITE** ▸*Math* Explain how you would use properties to write an expression equivalent to $7y + 4b - 3y$.

_____

_____

_____

Lesson Check (6.EE.A.3)

1. A ticket to a museum costs $8. A ticket to the dinosaur exhibit costs $5. The expression $8n + 5n$ represents the cost in dollars for n people to visit the museum and the exhibit. What is a simpler form of the expression $8n + 5n$?

2. What is an expression that is equivalent to $3(2p - 3)$?

Spiral Review (6.RP.A.3c, 6.NS.B.2, 6.EE.A.2b, 6.EE.A.3)

3. A Mexican restaurant received 60 take-out orders. The manager found that 60% of the orders were for tacos and 25% of the orders were for burritos. How many orders were for other items?

4. The area of a rectangular field is 1,710 square feet. The length of the field is 45 feet. What is the width of the field?

5. How many terms are in $2 + 4x + 7y$?

6. Boxes of cereal usually cost $4, but they are on sale for $1 off. A gallon of milk costs $3. The expression $4b - 1b + 3$ can be used to find the cost in dollars of buying b boxes of cereal and a gallon of milk. Write the expression in simpler form.

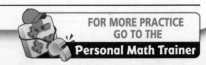

FOR MORE PRACTICE
GO TO THE
Personal Math Trainer

Name _____

Identify Equivalent Expressions

Essential Question How can you identify equivalent algebraic expressions?

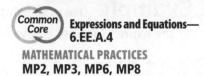

Common Core
Expressions and Equations—
6.EE.A.4
MATHEMATICAL PRACTICES
MP2, MP3, MP6, MP8

Unlock the Problem

Each train on a roller coaster has 10 cars, and each car can hold 4 riders. The expression $10t \times 4$ can be used to find the greatest number of riders when there are t trains on the track. Is this expression equivalent to $14t$? Use properties of operations to support your answer.

• What is one property of operations that you could use to write an expression equivalent to $10t \times 4$?

_____

_____

 Determine whether $10t \times 4$ is equivalent to $14t$.

The expression $14t$ is the product of a number and a variable, so rewrite $10t \times 4$ as a product of a number and a variable.

Use the Commutative Property of Multiplication.

$10t \times 4 = 4 \times$ _____

Use the _____
Property of Multiplication.

$= (4 \times$ _____$) \times t$

Multiply within the parentheses.

$=$ _____

Compare the expressions $40t$ and $14t$.

Think: 40 times a number is not equal to 14 times the number, except when the number is 0.

Check by choosing a value for t and evaluating $40t$ and $14t$.

Write the expressions.	$40t$	$14t$
Use 2 as a value for t.	$40 \times$ _____	$14 \times$ _____
Multiply. The expressions have different values.	_____	_____

So, the expressions $10t \times 4$ and $14t$ are _____.

Math Talk

MATHEMATICAL PRACTICES ⑥

Explain Why are the expressions $7a$ and $9a$ not equivalent, even though they have the same value when $a = 0$?

Chapter 7 407

Example Use properties of operations to determine whether the expressions are equivalent.

Ⓐ $7y + (x + 3y)$ and $10y + x$

The expression $10y + x$ is a sum of two terms, so rewrite $7y + (x + 3y)$ as a sum of two terms.

Use the Commutative Property of Addition to rewrite $x + 3y$.

$$7y + (x + 3y) = 7y + (\_\_\_\_\_ + \_\_\_\_\_)$$

Use the _____ Property of Addition to group like terms.

$$= (\_\_\_\_\_ + 3y) + x$$

Combine like terms.

$$= \_\_\_\_\_ + x$$

Compare the expressions $10y + x$ and $10y + x$: They are the same.

So, the expressions $7y + (x + 3y)$ and $10y + x$

are _____ .

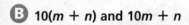 **Math Talk**

MATHEMATICAL PRACTICES ⑧

Generalize Explain how you can decide whether two algebraic expressions are equivalent.

Ⓑ $10(m + n)$ and $10m + n$

The expression $10m + n$ is a sum of two terms, so rewrite $10(m + n)$ as a sum of two terms.

Use the Distributive Property.

$$10(m + n) = (10 \times \_\_\_\_\_) + (10 \times \_\_\_\_\_)$$

Multiply within the parentheses.

$$= 10m + \_\_\_\_\_$$

Compare the expressions $10m + 10n$ and $10m + n$.

Think: The first terms of both expressions are _____ , but the second terms are different.

Check by choosing values for m and n and evaluating $10m + 10n$ and $10m + n$.

Write the expressions.	$10m + 10n$	$10m + n$
Use 2 as a value for m and 4 as a value for n.	$10 \times \_\_\_\_\_ + 10 \times \_\_\_\_\_$	$10 \times \_\_\_\_\_ + \_\_\_\_\_$
Multiply.	$\_\_\_\_\_ + \_\_\_\_\_$	$\_\_\_\_\_ + \_\_\_\_\_$
Add. The expressions have different values.	_____	_____

So, the expressions $10(m + n)$ and $10m + n$ are

_____ .

 Math Talk

MATHEMATICAL PRACTICES ③

Apply How do you know that the terms $10n$ and n from Part B are not equivalent?

Name _____

Share and Show MATH BOARD

Use properties of operations to determine whether the expressions are equivalent.

1. $7k + 4 + 2k$ and $4 + 9k$

Rewrite $7k + 4 + 2k$. Use the Commutative Property of Addition.

$$7k + 4 + 2k = 4 + \underline{\hspace{1cm}} + 2k$$

Use the Associative Property of Addition.

$$= 4 + (\underline{\hspace{1cm}} + \underline{\hspace{1cm}})$$

Add like terms.

$$= 4 + \underline{\hspace{1cm}}$$

The expressions $7k + 4 + 2k$ and $4 + 9k$ are _____.

2. $9a \times 3$ and $12a$

3. $8p + 0$ and $8p \times 0$

4. $5(a + b)$ and $(5a + 2b) + 3b$

Math Talk

MATHEMATICAL PRACTICES ②

Reasoning How do you know that $x + 5$ is not equivalent to $x + 8$?

On Your Own

Use properties of operations to determine whether the expressions are equivalent.

5. $3(v + 2) + 7v$ and $16v$

6. $14h + (17 + 11h)$ and $25h + 17$

7. $4b \times 7$ and $28b$

8. **GO DEEPER** Each case of dog food contains c cans. Each case of cat food contains 12 cans. Four students wrote the expressions below for the number of cans in 6 cases of dog food and 1 case of cat food. Which of the expressions are correct?

$6c + 12$ $\qquad$ $6c \times 12$ $\qquad$ $6(c + 2)$ $\qquad$ $(2c + 4) \times 3$

Problem Solving • Applications (Real World)

Use the table for 9–11.

Collectible Cards	
Type	**Number per Packet**
Baseball	b
Cartoon	c
Movie	m
Animal	a

9. Marcus bought 4 packets of baseball cards and 4 packets of animal cards. Write an algebraic expression for the total number of cards Marcus bought.

_____

10. **MATHEMATICAL PRACTICE ③ Make Arguments** Is the expression for the number of cards Marcus bought equivalent to $4(a + b)$? Justify your answer.

_____

_____

_____

_____

WRITE ▸ *Math* • **Show Your Work**

11. **THINK SMARTER** Angelica buys 3 packets of movie cards and 6 packets of cartoon cards and adds these to the 3 packets of movie cards she already has. Write three equivalent algebraic expressions for the number of cards Angelica has now.

_____

_____

12. **THINK SMARTER** Select the expressions that are equivalent to $3(x + 2)$. Mark all that apply.

Ⓐ $3x + 6$

Ⓑ $3x + 2$

Ⓒ $5x$

Ⓓ $x + 5$

Identify Equivalent Expressions

Use properties of operations to determine whether the expressions are equivalent.

Common Core

COMMON CORE STANDARD—6.EE.A.4
Apply and extend previous understandings of arithmetic to algebraic expressions.

1. $2s + 13 + 15s$ and
$17s + 13$

_____ equivalent _____

2. $5 \times 7h$ and $35h$

_____

3. $10 + 8v - 3v$ and $18 - 3v$

_____

4. $(9w \times 0) - 12$ and $9w - 12$

_____

5. $11(p + q)$ and
$11p + (7q + 4q)$

_____

6. $6(4b + 3d)$ and $24b + 3d$

_____

7. $14m + 9 - 6m$ and $8m + 9$

_____

8. $(y \times 1) + 2$ and $y + 2$

_____

9. $4 + 5(6t + 1)$ and $9 + 30t$

_____

10. $9x + 0 + 10x$ and $19x + 1$

_____

11. $12c - 3c$ and $3(4c - 1)$

_____

12. $6a \times 4$ and $24a$

_____

Problem Solving · Real World

13. Rachel needs to write 3 book reports with b pages and 3 science reports with s pages during the school year. Write an algebraic expression for the total number of pages Rachel will need to write.

_____

14. Rachel's friend Yassi has to write $3(b + s)$ pages for reports. Use properties of operations to determine whether this expression is equivalent to the expression for the number of pages Rachel has to write.

_____

15. **WRITE** ▸ *Math* Use properties of operations to show whether $7y + 7b + 3y$ and $7(y + b) + 3b$ are equivalent expressions. Explain your reasoning.

_____
_____

Lesson Check (6.EE.A.4)

1. Ian had 4 cases of comic books and 6 adventure books. Each case holds c comic books. He gave 1 case of comic books to his friend. Write an expression that gives the total number of books Ian has left.

2. In May, Xia made 5 flower planters with f flowers in each planter. In June, she made 8 flower planters with f flowers in each planter. Write an expression in simplest form that gives the number of flowers Xia has in the planters.

Spiral Review (6.RP.A.3c, 6.RP.A.3d, 6.EE.A.2c, 6.EE.A.3)

3. Keisha wants to read for 90 minutes. So far, she has read 30% of her goal. How much longer does she need to read to reach her goal?

4. Marvyn travels 105 miles on his scooter. He travels for 3 hours. What is his average speed?

5. The expression $5(F - 32) \div 9$ gives the Celsius temperature for a Fahrenheit temperature of F degrees. The noon Fahrenheit temperature in Centerville was 86 degrees. What was the temperature in degrees Celsius?

6. At the library book sale, hardcover books sell for $4 and paperbacks sell for $2. The expression $4b + 2b$ represents the total cost for b hardcover books and b paperbacks. Write a simpler expression that is equivalent to $4b + 2b$.

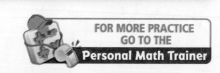

FOR MORE PRACTICE
GO TO THE
Personal Math Trainer

Name _____

Personal Math Trainer
Online Assessment and Intervention

1. Use exponents to rewrite the expression.

$3 \times 3 \times 3 \times 3 \times 5 \times 5$

☐ ☐
3   × 5

2. A plumber charges $10 for transportation and $55 per hour for repairs. Write an expression that can be used to find the cost in dollars for a repair that takes h hours.

_____

3. Ellen is 2 years older than her brother Luke. Let k represent Luke's age. Identify the expression that can be used to find Ellen's age.

Ⓐ $k - 2$

Ⓑ $k + 2$

Ⓒ $2k$

Ⓓ $\frac{k}{2}$

4. Write 4^3 using repeated multiplication. Then find the value of 4^3.

_____

5. Jasmine is buying beans. She bought r pounds of red beans that cost $3 per pound and b pounds of black beans that cost $2 per pound. The total amount of her purchase is given by the expression $3r + 2b$. Select the terms of the expression. Mark all that apply.

Ⓐ 2

Ⓑ $2b$

Ⓒ 3

Ⓓ $3r$

Assessment Options
Chapter Test

6. Choose the number that makes the sentence true.

The formula $V = s^3$ gives the volume V of a cube with side length s.

The volume of a cube that has a side length of 8 inches

is | 24 | inches cubed.
| 64 |
| 512 |

7. Liang is ordering new chairs and cushions for his dining room table. A new chair costs $88 and a new cushion costs $12. Shipping costs $34. The expression $88c + 12c + 34$ gives the total cost for buying c sets of chairs and cushions. Simplify the expression by combining like terms.

_____

8. Mr. Ruiz writes the expression $5 \times (2 + 1)^2 \div 3$ on the board. Chelsea says the first step is to evaluate 1^2. Explain Chelsea's mistake. Then, evaluate the expression.

9. Jake writes this word expression.

the product of 7 and m

Write an algebraic expression for the word expression. Then, evaluate the expression for $m = 4$. Show your work.

414

10. Sora has some bags that each contain 12 potatoes. She takes 3 potatoes from each bag. The expression $12p - 3p$ represents the number of potatoes p left in the bags. Simplify the expression by combining like terms. Draw a line to match the expression with the simplified expression.

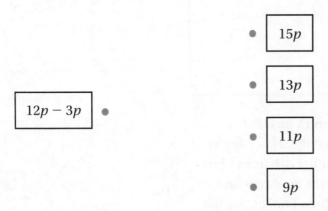

$15p$

$13p$

$12p - 3p$ •

$11p$

$9p$

11. **GO DEEPER** Logan works at a florist. He earns $600 per week plus $5 for each floral arrangement he delivers. Write an expression that gives the amount in dollars that Logan earns for delivering f floral arrangements. Use the expression to find the amount Logan will earn if he delivers 45 floral arrangements in one week. Show your work.

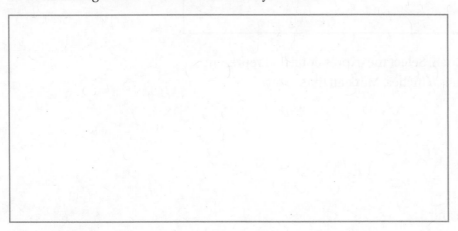

12. Choose the word that makes the sentence true.
Dara wrote the expression $7 \times (d + 4)$ in her notebook. She used the

Associative
Commutative
Distributive

Property to write the equivalent expression $7d + 28$.

13. Use properties of operations to determine whether $5(n + 1) + 2n$
and $7n + 1$ are equivalent expressions.

14. Alisha buys 5 boxes of peanut butter granola bars and 5 boxes of
cinnamon granola bars. Let p represent the number of bars in a box of
peanut butter granola bars and c represent the number of bars in a box
of cinnamon granola bars. Jaira and Emma each write an expression
that represents the total number of granola bars Alisha bought. Are the
expressions equivalent? Justify your answer.

Jaira	Emma
$5p + 5c$	$5(p + c)$

15. Abe is 3 inches taller than Chen. Select the expressions that represent
Abe's height if Chen's height is h inches. Mark all that apply.

○ $h - 3$

○ $h + 3$

○ the sum of h and 3

○ the difference between h and 3

16. Write the algebraic expression in the box that shows an equivalent
expression.

| $3(k + 2)$ | $3k + 2k$ | $2 + 6k + 3$ |

$6k + 5$	$5k$	$3k + 6$

17. Draw a line to match the property with the expression that shows the property.

Associative Property of Addition •   • $0 + 14 = 14$

Commutative Property of Addition •   • $14 + b = b + 14$

Identity Property of Addition •   • $6 + (8 + b) = (6 + 8) + b$

Personal Math Trainer

18. THINK SMARTER + A bike rental company charges \$10 to rent a bike plus \$2 for each hour the bike is rented. An expression for the total cost of renting a bike for h hours is $10 + 2h$. Complete the table to find the total cost of renting a bike for h hours.

Number of Hours, h	$10 + 2h$	Total Cost
1	$10 + 2 \times 1$	
2		
3		
4		

19. An online sporting goods store charges \$12 for a pair of athletic socks. Shipping is \$2 per order.

Part A

Write an expression that Hana can use to find the total cost in dollars for ordering n pairs of socks.

_____

Part B

Hana orders 3 pairs of athletic socks and her friend, Charlie, orders 2 pairs of athletic socks. What is the total cost, including shipping, for both orders? Show your work.

20. Fernando simplifies the expression $(6 + 2)^2 - 4 \times 3$.

Part A

Fernando shows his work on the board. Use numbers and words to explain his mistake.

$(6 + 2)^2 - 4 \times 3$

$(6 + 4) - 4 \times 3$

$10 - 4 \times 3$

6×3

18

Part B

Simplify the expression $(6 + 2)^2 - 4 \times 3$ using the order of operations.